RAINFOREST AUSTRALIA

Rainforest Creek, Mount Lewis, North Queensland

Myrtle Beech (Nothofagus cunninghamii), *Tasmania*

Antarctic Beech (Nothofagus moorei), *Lamington National Park, SE Queensland*

FORESTS PRIMEVAL

To walk in rainforest is to be carried far back in time. The distant ancestors of the trees which surround you may have spread their leaves to the sunshine when dinosaurs still roamed the earth. When its rainforests first came into being, around 100 million years ago, Australia was part of the supercontinent Gondwana, an enormous land mass which also included Africa, South America, India and Antarctica. When, about 65 million years ago, catastrophic asteroid impacts created a long "winter" which destroyed most of the Northern Hemisphere's rainforests, the Gondwanan forests remained intact. Gradually Gondwana broke into pieces. Australia finally became independent around 45 million years ago, when it parted from Antarctica and drifted slowly northwards. As the new continent grew drier, the rainforests which covered it shrank until they survived only in the north and east. Human activities, especially in the past 200 years, have hastened rainforest decline. Today, Australia's rainforests cover just three-thousandths of the continent's surface, and are fragmented into thousands of remnants.

Strangler fig, Mount Tamborine, SE Queensland

Josephine Falls, Wooroonooran National Park, North Queensland

Opposite: "The Ballroom", Cradle Mountain-Lake Dove National Park, Tasmania

5

RAIN, THE LIFEBLOOD OF THE FOREST

Look at rainforest from some vantage point — say the window of an aircraft flying past an isolated stand, or from a neighbouring mountain top — and you will see it wears cloud, wrapped about it like a cloak, or hovering above it like a halo. This cloud, which shields the forest from the full force of the sun's rays, is made up of minute particles of moisture. When the air temperature falls, the droplets clump together, form raindrops and fall.

For best growth, a rainforest needs to receive at least 1300 millimetres of life-giving rain each year, spread evenly throughout the twelve months. One-quarter is used by trees, bushes, creepers, vines and other plants, whose roots suck it from the soil and the air. Three-quarters returns to the air, evaporated by the sun's warmth or breathed out by leaves, to float above the forest as cloud until cooler conditions cause it to fall again as rain.

Left: Dorrigo National Park, New South Wales
Inset far left, left and above: Russell Falls, Mount Field National Park, Tasmania;
Wooroonooran National Park, North Queensland;
Raindrops on palm leaf

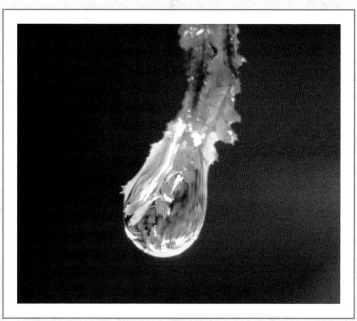

Raindrop

7

ADVENTURE IN THE MIST

The cloud which floats above rainforest sinks at night, so that morning finds the trees enveloped in mist. Enter such a forest at dawn and you will stand in a bubble of clear, moist air. The tree trunks near you can be clearly seen — every filament of moss jewelled with a tiny diamond of dew, the lichens silver-filmed with moisture. The trees ten metres away are less clearly defined, and at twenty metres the trunks are mere outlines in the pearly mist.

In this new and hazy world, sounds become magnified. In the canopy far above, pigeons are breakfasting, their presence betrayed by muffled "oom"s and by fruit pattering to the ground. The moist leaf litter muffles your footsteps, so that some small creature is taken unaware and scuttles away in a flash of brown fur into a hidey-hole at the base of a decaying forest giant, where the roots of a strangler fig touch ground. A whipbird calls at ear-shattering volume, its mate completing the territorial warning with two ringing, defiant notes.

For the explorer, mist gives the rainforest added allure. For the forest it is vital, bathing it in soothing moisture and renewing the vitality of its most secret places.

Left and above: Wooroonooran National Park, North Queensland

Crows Nest Fern in subtropical rainforest

SUNLIGHT POWERS A GREEN MACHINE

unlight often reaches the rainforest with its force softened by
assage through cloud, but the sun's rays are necessary for the
nctioning of all green plants. Within a plant's tissues, especially
its leaves, are tiny green particles which are powered by
nlight to turn minerals and water into food substances which
e plant can use for growth and maintenance.

order to expose its leaves to sunlight, a rainforest tree must
ow upwards until its crown takes its place in the forest canopy.
he bird flying over rainforest searching for a flowering or
uiting tree on which to feed looks down on a jigsaw of
terlocking crowns, each tree's leaves ending just before its
eighbour's leaves begin. This closed canopy prevents much
nlight penetrating to the ground, and smaller plants must have
eir own strategies for gaining a place in the sun. Some, like
rns, orchids and mosses, are carried upwards as they cling to
e trunks of other plants.

pposite: Eungella National Park, North Queensland

Moss, Bunya Mountains, Queensland

11

LAMENT FOR LOST FORESTS

We've mimicked forests,
planted smooth dark trees
row upon row,
reaching side by side together
towards a distant sun.
Gone the whipbird's echo
and the rosella's chime,
now there are just planted cages
of silence.

Where once the silver-voiced lyrebird
had songs to mimic,
poured out his heart
in exultant cascades,
now we hear barking dogs,
sawing of timber,
a car that will not start —
a sad impersonation.

Steve Parish

FROM THE CANOPY TO THE GROUND

The canopy of one of Australia's tropical or subtropical rainforests consists of the crowns of up to sixty species of trees, whose trunks rise branchless for up to thirty metres above their flaring, supportive buttress roots. Some of these trees may drop their leaves at a certain time each year. Some may produce flowers and fruit every year, or only every three or four years. The canopy provides food for many animals. During the day, birds such as parrots and pigeons and hordes of insects seek nectar and fruit; at night, the canopy is alive with possums, blossom-eating bats, frogs and insects.

The flowers and fruit of a tropical rainforest tree may grow direct from the trunk amongst orchids and ferns, lichens and mosses. Climbing palms, vines and strangler figs clamber up tall trunks and smaller trees, while shrubs form a lower storey of plant life.

From the moist ground grow ferns and large-leaved plants, and fallen logs and leaf-litter are crusted with fungi of varying shapes and colours. The ground floor of the forest is home to many animals, from birds such as the jewelled pittas and avenue-building bowerbirds to a variety of mammals, reptiles, frogs and smaller creatures. All that lives in a rainforest eventually falls to its floor, where insects, fungi and time begin the process of returning its substance to the trees once again.

Opposite: Crows Nest Fern on vine, Lamington National Park, southeastern Queensland

Green Possum

Day-flying Zodiac Moth

STANLEY BREEDEN

Red-eyed Tree-frogs

Ground litter in a rainstorm

Noisy Pitta

COLIN LLOYD

ENTER THE GREEN GALLERY

Both palms and ferns are found in profusion in Australia's tropical and subtropical rainforests, while in temperate rainforests palms are absent but ferns come into their own. These two types of plants have long been regarded as particular glories of rainforest. In his book *Excursions in New South Wales*, published in 1833, William Breton wrote about the Illawarra rainforest south of Sydney:

> The palms, from fifty to eighty feet high, and quite straight, the fern trees, parasitical plants, and climbers, were beautiful and in many places so luxuriant was the vegetation, and so completely were the climbers, many of them nearly as large as a man's body, interwoven amongst trees, that they rendered the forest off the path, utterly impervious.

In an article in a series titled *Excursions from Berrima* published in the 1860s, novelist and amateur botanist Louisa Atkinson described a rainforest as follows:

> The nature of the country had now changed to a dense semi-tropical thicket; while the sun's rays were obscured by the overhanging branches, or fell like mosaic upon the moss-green stones and ferns. Cabbage-palms stretched their slender stems above the tangled copse...the tree-ferns, elk horn and bird's nest ferns revelled in the humid shade.

The popularity of palms and ferns has made many species commonplace as garden plants. However, no garden viewing can compare with the thrill of seeing a stand of graceful Fan Palms in a northeastern Queensland tropical setting, or walking amongst tall tree ferns in the cool air of the mountain gullies of Victoria or Tasmania.

Far left: Fan Palms, Daintree River National Park, North Queensland

Entering the green gallery

Red Kurrajong flowers attract nectar-seeking birds

Flowers of the Pink Silky Oak and a Macleay's Honeyeater

STANLEY BREEDEN

Flower of the Powderpuff Lillypilly

STANLEY BREEDEN

Tick Orchids

A FESTIVAL OF FLOWERS

A rainforest tree may not blossom until it is several decades old, then may flower only every three or four years thereafter. However, when the time comes, such a tree bursts forth in a spectacular massed display. Some tropical rainforest trees bear abundant flowers and fruit on their trunks and larger branches, a phenomenon known as cauliflory.

The "greenhouse" conditions of constant humidity and temperature which rule under the canopy favour the growth of fragile or fleshy flowers such as orchids.

Rainforest flowers advertise their presence to nectar-seeking birds, bats, possums and insects by attractive colours or, in some cases, delicious scents. Red petals are particularly attractive to day-flying birds, while white, pale pink or mauve blossoms lure night-flying bats, and insects such as moths. Of course the flower does not provide nectar simply for the good of its animal guests. Each diner leaves with grains of pollen adhering to its body, then carries them to another flower and deposits them there, hopefully fertilising the waiting germ cells to form seeds. A number of Australian animals, for example the birds called honeyeaters, regularly act as pollen couriers in the course of obtaining their daily meals.

Above: Spider Orchid; above right: Northern Silky Oak

Above: Johnstone River Satinash; above right: Cooktown Orchid

Above: Sprouting Bumpy Satinash; above right: Bumpy Satinash

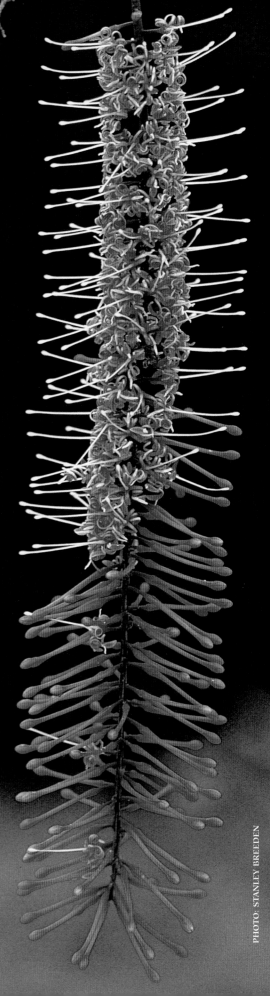

STANLEY BREEDEN

Above left: Rusty Fig; above: Native Peppe

STANLEY BREEDEN

Above left: Topaz Tamarind; above: Blue Nu

STANLEY BREEDEN

Above left: Atherton Oak; above: Grey Carrabee

Northern White Beech

Porcelain Fruit

Lethedon setosa

FRUITS OF THE FOREST

Rainforest plants have various ways of ensuring that their seeds reach new ground far from the parent plant. Some have pods which, when the seeds are ripe, spring open, propelling them into space. Others are covered by hooks or spines, which will attach to an animal's fur or feathers and be carried away. Many are encased in edible fruit, whose devourer will eventually expel the seeds surrounded by fertiliser. Rainforest fruits may contain antibiotics which protect the seeds against bacteria, or substances which repel animals unsuitable for seed dispersal.

Cassowary chicks eating fruit

THE CYCLE OF LIFE

A fungus is a plant, but does not contain the particles of chlorophyll which enable green plants to make food using the energy of sunlight, so it feeds by breaking down the tissues of plants and animals. The fruiting body of a fungus may take the shape of a toadstool, bracket, puffball, mushroom or cup. However it is small compared to the network of threads, or hyphae, which it sends into its food source.

Rainforest fungi grow amongst the ground litter, on dead wood and on tree trunks. They are nature's recyclers, which help return once-living matter to the soil to be used again. Some fungi even work in partnership with living plants — the fungus grows on the plant's roots, feeding on processed food excess to the plant's requirements. The plant in return uses raw nutrients gathered by the fungus's vast network of hyphae. This alliance allows trees to grow on soil they would otherwise find too poor.

Opposite: Mount Field National Park, Tasmania

Above: Luminous fungus; below: Cup fungus

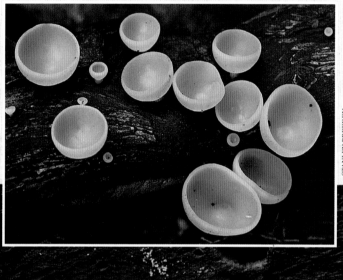

Fungi such as these tiny toadstools break down leaf litter and dead wood

Southern Cassowary

RAINFOREST GIANT

The Southern Cassowary is found in small, isolated populations in rainforests from Cape York to Townsville. A feeder upon fallen fruit, it soon vanishes from an area where its habitat is disturbed. Australia's second-tallest but heaviest bird is flightless, plumaged in glossy black, double-shafted feathers and crowned with a horny casque. The female courts the smaller male, lays four or five eggs, then leaves him to incubate them. He will take care of the striped chicks for at least four months.

Southern Cassowary chicks and adult

RAINFOREST EXPERIENCE

The forest stands guard around us,
calm and still.
Before us, the waterfall
spills in silver splendour
over fern-draped terraces.

We share
the songs of the forest,
in a closeness of spirit
that needs no words.
Together we dream, hope, imagine.

In the forest each of us finds our heart's desire.

WHERE RAINFOREST AND REEF MEET

The area from Cooktown to Paluma, Queensland, surely one of the most beautiful places on earth, was given World Heritage listing in 1988. Here, rainforest-clad, cloud-capped mountains descend to the sparkling Pacific Ocean, whose translucent sapphire waters harbour another of the world's natural miracles, the Great Barrier Reef.

The Daintree River National Park extends from the Bloomfield River south to the Mossman River. In 1873, George Elphinstone Dalrymple journeyed by whaleboat up the Daintree River, which he named after a geologist friend. He wrote of the Daintree that:

> ...no river-reach in North Australia possesses surroundings combining so much of distant mountain grandeur with local beauty and wealth of vegetation.

Left: Female Eclectus Parrot

Cape Tribulation National Park, North Queensland; opposite: Cape Tribulation National Park, North Queensland

Pittosporum *sp.*

Boyd's Forest Dragon

Mossman River, Daintree National Park, North Queensland

Hypipamee Crater, Atherton Tableland, North Queensland

Herbert River Ringtail Possum

Ulysses Butterfly

STANLEY BREEDEN

STANLEY BREEDEN

32

Wanggoolba Creek, Fraser Island, Queensland

WORLD HERITAGE RAINFOREST PARKS

World Heritage listing is a recognition by the international community
that an area is the best of its kind in the world, possessing exceptional
natural and/or cultural values. Rainforest is a major constituent of
three of Australia's World Heritage listed areas. They are the Wet Tropics
area of Queensland (inscribed 1988), the Central Eastern Rainforests of Australia,
New South Wales and Queensland (inscribed 1986, extended 1994), the Tasmanian Wilderness
area in the island State's southwest (inscribed 1982, extended 1989) and Fraser Island, off the
coast of Queensland (inscribed 1992), whose unique rainforest grows on sand. Another World
Heritage area, Kakadu National Park, Northern Territory (inscribed 1981 and extended 1987 and
1992), also contains fragments of rainforest.

The Land Mullet is a large, ground-living lizard

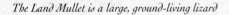

Brown Tree-snake dining

ENCOUNTERS OF THE SCALED KIND

One of the great thrills of rainforest exploration is to spot a snake or lizard. It may be sunning itself in the sunshine which shafts down through the gap where a tree has fallen, or it may be crawling across a walking track. Keep a watch out at eye level too, for pythons like to coil up in tree forks or in the hearts of ferns, and many lizards spend most of their lives on tree trunks or among the foliage. On a lichen-crusted trunk, you may discover a forest dragon, mailed like a knight of old, with a smart crest of scales adorning its head and spine.

On the forest floor, or basking on a boulder, you may see the shiny-scaled Land Mullet, a lizard which can grow to nearly 30 centimetres in length. It lives with others of its kind, foraging for fallen fruit, fungi, snails, slugs and insects, and shelters in hollow logs and under buttress roots. Night-time is gecko time, when velvety-skinned, huge-eyed little hunters roam the trees, pouncing on moths, spiders and other unwary prey. Geckos have no eyelids, and every so often you will see them pause to wipe their eyes clean with their moist pink tongues.

If you do not want to see snakes, tramp along, talking loudly. If you do, walk softly and watch ahead for a slither of movement. Any sensible snake will move away as soon as it senses your presence, but you may catch a glimpse of one of Nature's most fascinating creatures, perfectly adapted to its armless, legless life.

Ring-tailed Gecko

Green Python

Southern Forest Dragon

Fringed Tree-frog

Roth's Tree-frog

Water Frog

Green-thighed Frog

Stream Frog   *Fleay's Frog*

VANISHING RAINFOREST TREASURES

The frogs of Australia's rainforests are living marvels — beautiful creatures, with fascinating and often amazing lifestyles. The rainforests provide ideal frog habitats, for the high humidity of air prevents thin, moist skins from drying out. Night is the best time to explore the forest, torch in hand, tracking down the croakings and chirpings that signal male frogs calling to attract prospective mates.

Rainforest is so humid that here some species manage to break froggy tradition and lay their eggs out of water, on the undersides of leaves or in the leaf litter. The soft globes, in their masses of egg jelly, remain so moist that tadpoles can develop into tiny frogs inside their own fluid-filled life-support capsules. Other rainforest frogs spawn in constantly renewed puddles of water lying in tree hollows, or in pools caught in epiphytic plants. Some frogs which live on the rainforest floor guard their eggs until they hatch. The male Marsupial Frog has "hip-pockets" into which his tadpoles slither; they stay in these pouches until ready to emerge as miniature adults.

Most remarkable are the two gastric-brooding frogs, whose females swallow their own eggs, which hatch in their stomachs. While the tadpoles are changing into frogs, their mother does not eat. The babies finally emerge through her opened mouth as tiny, perfectly formed frogs.

White-lipped Tree-frog

STANLEY BREEDEN

Sharp-nosed Torrent-frog

STANLEY BREEDEN

Northern Barred Frog

FROGS GIVE EARLY WARNING

Frogs, with their sensitive skins, continual need for moisture and easily damaged eggs, are extremely sensitive to changes in their habitat. In fact, they form an environmental "early warning system", since their disappearance from an area may indicate problems with air, soil, surface water or rainfall quality. Since the late 1970s, frogs have declined in numbers worldwide. Some reasons given are global warming, pollution of wetlands by chemicals and salt, habitat loss, and increasing ultra-violet radiation as the ozone layer of the atmosphere thins.

At present, at least thirty-four of Australia's frog species are endangered, or at risk if their habitat should change. Rainforest frogs are particularly vulnerable, because of the limited extent of their habitat and their specialised requirements. In the past two decades, a number of rainforest frogs, including the two gastric-brooding frogs, have apparently disappeared from mountain rainforest streams between Brisbane and Cairns, in Queensland, and other species are becoming rare.

The Australian National Conservation Agency has initiated a plan to identify frog species requiring conservation and there are studies being made of the declining frogs of Queensland and New South Wales.

Leaf-green Tree-frog

Red-eyed Tree-frog

White-lipped Tree-frog

Above left: Male Eclectus Parrot; above right: Female Yellow-bellied Sunbird

Above left: Australian King-Parrot; above right: Male Lovely Fairy-wren

Above left: Crimson Rosella; above right: Male Golden Whistler

BIRDS OF THE RAINFOREST

Australia's northern rainforests are home to a great number of birds of many different kinds. This abundance and diversity of feathered creatures is possible because the rainforest offers a wide variety of places in which to feed. In the canopy, flower nectar, fruit and the insects which they attract are eaten by honeyeaters, bowerbirds, pigeons and parrots. The tree trunks, vines, creepers and lower storey of smaller trees and bushes harbour insects, spiders, lizards, frogs and other food for songbirds such as wrens, robins, flycatchers and fantails. The ground litter is rich in small crawling creatures, including worms, snails, centipedes and insects, and forms a dining-place for scrub-robins, scrub-birds, pittas, lyrebirds, kingfishers and other birds which scratch up, or swoop down upon, their prey.

Red-bellied Pitta

GRAEME CHAPMAN

Buff-breasted Paradise-Kingfisher

STANLEY BREEDEN

Green Catbird

MICHAEL MORCOMBE

41

Eastern Yellow Robin on its lichen-decorated nest

Female Shining Flycatcher on nest

Rufous Fantail with small chicks

NURSERIES FOR SONGBIRDS

The birds of the rainforest take great care in selecting their nest sites and then in concealing the nests which they build.

A pair of Eastern Yellow Robins will choose a sturdy tree fork, then construct in it a deep cup made from bark, grass and other fibre cemented together with cobwebs. Lichen is added to the outside to camouflage the nest. The Grey-headed Robin builds its nest on a horizontal branch, perhaps choosing a site on a spiky lawyer-vine for added protection. Flycatchers construct neat little cups on flimsy twigs or vines, then add lichen to the outside walls. Fantails make neat fibre bowls, often with tails hanging underneath. The Eastern Whipbird places its nest deep in a thicket and decorates it with mosses.

Predators roam the forest day and night, and possums, snakes, quolls and lizards take many eggs and nestlings. However, each year enough feathered songsters emerge from their cradles to make the dawn chorus in a rainforest a memorable experience.

Eastern Whipbird feeding chicks

Grey-headed Robin on nest

Pied Imperial-Pigeon

Wonga Pigeon

Topknot Pigeon

Elegant Imperial-Pigeon

PIGEONS AND DOVES

The rocketing take-off and swift, wing-pumping flight of a pigeon or dove bear witness to the massive pectoral muscles which power its wings and give this group of birds their full-breasted shapes. They are capable of flying great distances, and some rainforest pigeons and doves roam widely while searching for the fruits on which they feed. The Pied Imperial-Pigeon migrates to New Guinea each year, returning to nest on islands off the coast of northeastern Australia. It must fly to the mainland each day to feed on rainforest fruits, then return to the islands late each afternoon to feed its young and to roost. Pigeons and doves lay two white eggs on their flimsy stick nests.

Emerald Dove

RAOUL SLATER

Male Golden Bowerbird on his display stage

Male Regent Bowerbird in his avenue bower

BOWERBIRDS

Male bowerbirds spend most of the breeding season building and decorating their bowers, special places where they sing and "dance" to show off to females. Most Australian bowerbirds build avenues of sticks, which they may paint with saliva and crushed plants, but the Golden Bowerbird makes a display stage between two tree trunks. If a female is impressed by the male's actions, his bower and the treasures he has collected, she mates with him. Then she builds a nest and incubates her eggs and feeds her chicks alone.

Male Satin Bowerbird adding a stick to his avenue bower

45

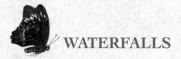

WATERFALLS

Where a rainforest stream runs across a bed of hard rock onto a bed of softer rock, the latter is worn away by the impact of water and by sand and rocks swirled around in the flow. Gradually, the fall deepens and a pool forms at its foot.

Waterfalls, surrounded by ferns, plunging in sheer cascades or veiling glistening rocks in crystalline beauty, provide some of the loveliest sights in Australia's rainforests. These scenic charms are usually well signposted and can be reached by walking along well-maintained tracks. Though the end of such a journey is a delight, walking through the forest is also very rewarding.

Left: Nandroya Falls, Wooroonooran National Park, North Queensland
Far right inset: Russell Falls, Mount Field National Park, Tasmania
Right inset: Josephine Falls, Wooroonooran National Park, North Queensland

Natural Arch, Springbrook National Park, southeastern Queensland

Elebana Falls, Lamington National Park, southeastern Queensland

RAINFOREST WATERWAYS

Rain falls upon the rainforest canopy, then plummets to the soil. Some is absorbed by vegetation, while the rest flows in a network of tiny streams into creeks, which eventually contribute their flow to rivers.

At altitude, rainforest streams are often cold and their water is charged with oxygen. They harbour hosts of insects, such as caddis-fly and dragonfly larvae, giant waterbugs, water scorpions and whirligig beetles, as well as tadpoles and worms. Spiny crayfish are found in fast-moving streams; in deeper water, and in pools where the flow is retarded, freshwater turtles and the Platypus make their homes.

The unique Platypus hunts its food of small water animals such as shrimps and insect larvae in clear, unpolluted water. It closes its eyes, ears and nostrils when it dives, detecting its prey by means of the tiny electrical signals their bodies betray to its sensitive, rubbery bill. Food is stored in cheek-pouches until the Platypus surfaces to eat. A female Platypus digs a nest-burrow in the bank of a stream and in a nursery chamber at its end lays two round, soft-shelled eggs. She keeps these warm between her tail and her body until they hatch, then feeds the babies on milk they suckle from her abdomen.

Left: Johnstone River, Wooroonooran National Park, North Queensland
Far left inset: Lamington Blue Cray
Left inset: Eastern Long-necked Turtle

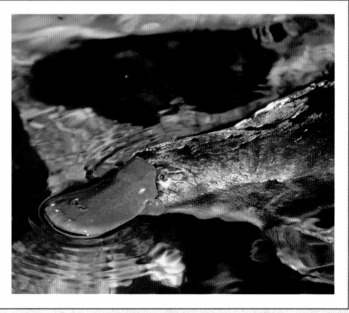

Platypus

Northern Ringtail Possum

Herbert River Ringtail Possum

Striped Possum

POSSUMS

Those who only enter the rainforest by day will miss seeing some of the most fascinating creatures which exist in its leafy galleries.

During daylight hours, possums are curled up in tree hollows, or in the crowns of ferns, or in soccer-ball sized nests they have woven of leaves and twigs, sleeping until dusk. The unusual Green Ringtail of North Queensland forests just hunkers down on a branch, tail curled beneath it and head between its hindlegs. There it sleeps soundly, camouflaged by its lichen-toned fur, which appears green, but is actually made up of black, yellow and white hairs.

As night falls, there is much stretching and yawning and grooming of soft fur and splendid whiskers before possum emerge to search the trees for leaves, blossoms and fruits. The agile, strong-smelling Striped Possum uses special elongated, slender fingers to probe into holes in wood for grubs, other insects and spiders, which it crunches with relish. The slow-climbing Spotted Cuscus, which is found only in rainforests on the tip of Cape York, eats fruit and leaves, but may add small animals to its diet.

Daintree River Ringtail Possum

Green Ringtail Possum

Coppery Brushtail Possum

Spotted Cuscus

Diadem Horseshoe-bat

Long-nosed Potoroo

Red-necked Wallaby and joey

Spectacled Flying-fox

Platypus

Northern Quoll

Red-legged Pademelon

Musky Rat-kangaroo

RAINFOREST MAMMALS

Many different sorts of mammals (warm-blooded animals whose skins grow hair or fur, and which feed their young ones on milk) live in Australia's rainforests. One of the most remarkable is North Queensland's Musky Rat-kangaroo, a tiny, scampering relative of the familiar Red Kangaroo. Another kangaroo relative, the Red-necked Wallaby, is a common forest plant-eater, but the cat-sized potoroos, which depend on thick undergrowth for shelter, are becoming rare. The largest meat-eating marsupial is the Tasmanian Devil, now confined to the island State. It finds much of its food by scavenging, and as a hunter it cannot compete with the spotted, long-tailed quolls, which roam the forest at night hunting frogs, birds and small mammals. Large nectar-eating bats, the flying-foxes and blossom-bats, wing their way above the canopy at night, seeking flowering trees. Smaller insect-eating bats zip through forest clearings, scooping up flying insects as they go. They spend daylight hours in tree hollows.

Tasmanian Devil

Jewel bugs

Cruiser Butterfly

Long-horned grasshopper

Stag beetle

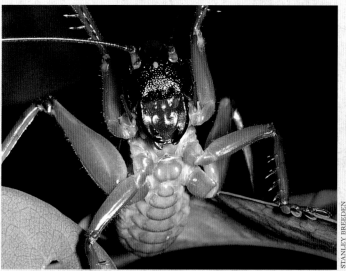

Cricket

Union Jack Butterflies newly emerged from their pupal cases

A SPLENDOUR OF INSECTS

t is easy to overlook some of the rainforest's most
orgeous inhabitants, for they hide under bark, or cling to
he undersides of leaves, or flit above the canopy, alighting
n flowers which they may rival in beauty. Rainforest
nsects are easiest to see in places where the canopy has
een disturbed, sunlight reaches the ground and there are
lenty of young plants, whose foliage is eagerly devoured.
ou will notice that some leaftips, which are coloured in
eds and bronzes, are unmarked — they contain natural
nsecticides to protect them from eager jaws. The forest
oor has its share of insects, often flightless, or reluctant
iers, which tend to gather under logs and litter during the
ay and to roam the open spaces at night. Look for these
round floor tenants on dull days, as they graze and hunt
mong the fallen leaves. Ground-dwelling insects are at
articular risk when rainforest burns or is felled, for they
ay find it impossible to move to another suitable patch of
rest, even if one exists nearby.

Mud-dauber wasp

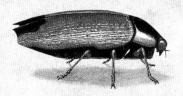

Leafwing Butterfly

Tropical rainforest, Daintree National Park, North Queensland

TOTAL AREA: 12 750 SQ KM
(NOT INCLUDING MIXED
RAINFORESTS WITH
EUCALYPTS)

CAPE
YORK
PEN

COOKTOWN

CAIRNS

N

TOWNSVILLE

QUEENSLAND

ROCKHAMPTON

BUNDABERG

FRASER
ISLAND

BRISBANE

NEW SOUTH WALES

QUEENSLAND

More than half of Australia's remaining rainforests, including tropical, subtropical, cool temperate and dry types, are found in Queensland. The largest single area of tropical rainforest, 7500 square kilometres in extent, occurs between Cooktown and Townsville, in the Wet Tropics of Queensland Tropical Heritage Area. At least 700 different sorts of plants and seventy sorts of birds, mammals and frogs occur only in the Wet Tropics.

Subtropical rainforest occurs in scattered remnants from the central Queensland coast to the New South Wales border, with major areas on the Conondale, Main and Macpherson Ranges. There is little more than 2000 squar kilometres of this forest, which has been heavily logged and isolated by clearing around it, so it is threatened by fire and weed invasion. The drier rainforests of the brigalow belt, which were dominated by Hoop and Buny. Pines, have mostly been cleared and today exist only in fragments in the Bunya Mountains and in Lamington National Park.

Dorrigo National Park, New South Wales

Tara-Bulga National Park, Strzelecki Ranges, Victoria

NEW SOUTH WALES

More than three-quarters of the rainforests that existed in New South Wales 200 years ago have been lost. The remnants are only partly protected and are vulnerable to fires which spread from neighbouring eucalypt forests.

Once, the subtropical rainforest known as the Big Scrub stretched over 750 square kilometres from Byron Bay to Lismore. Most has been cleared, as have the forests of the Illawarra, which were compared to South American jungle by early European explorers. Half of New South Wales' northeastern rainforests are included with southeastern Queensland forests in World Heritage listings. They preserve most of Australia's remaining Antarctic Beech and are home to around forty species of frogs and many species of birds, including primitive songbirds such as the Albert's Lyrebird and Rufous Scrub-bird. The State's southeastern forests include rainforest with tall eucalypts which provide homes for tree-living marsupials such as the Yellow-bellied and Feathertail Gliders. The Coolangubra forest is one of the few known refuges of the endangered Long-footed Potoroo.

VICTORIA

Ninety million years ago, the area that is now Victoria gave birth to new plant groups, which included the ancestors of eucalypts, banksias and waratahs. Today, Victoria's high rainfall plant communities are everywhere under threat and the rainforest which survives is in narrow stands less than ten hectares in area. The cool temperate forests of the Dandenong Ranges, the cloud forest of Wilsons Promontory and the Plumwood forests of the Howe Range are almost gone. Nearly eighty rainforest plant species are rare, or threatened with extinction.

The protection of Victoria's remaining rainforests, and of the mixed-forest areas where Mountain Ash, Alpine Ash and Shining Gum raise their splendid heads above the rainforest canopy, depends upon protection from logging and from accidental and deliberate burning. Clearfelling and fire allow eucalypts to regenerate, but destroy the more vulnerable rainforest trees. Leadbeater's Possum and the Long-footed Potoroo are only two endangered examples of the increasingly rare marsupials which will disappear if their habitat is obliterated.

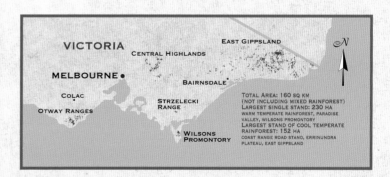

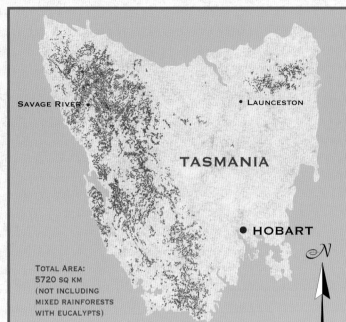

TASMANIA

Tasmania contains one-quarter of Australia's remaining rainforests, including the largest single tract of cool temperate rainforest in the world. Two-thirds of the island's rainforest is unprotected and it is the only State in Australia which still allows logging of pure rainforest. Tasmania's remaining forests include the last eighty square kilometre stand of the unique Huon Pine, and remnant forests of Pencil Pine and King Billy Pine, relatives of the Californian Redwood. Tasmania's rainforests are under threat from logging, mining, tourist access and fire. The largest area, in the State's northwest, has almost no conservation reserves and the integrity of the famous Tarkine Wilderness is being breached by a road. Half of the mixed rainforest/eucalypt forest which existed 200 years ago has already been cleared; more is targeted for sawn timber and woodchips. One-third of Tasmanian rainforest is included in National Parks. Its quality is shown by the fact that almost all is World Heritage listed.

Russell Falls, Mount Field National Park, Tasmania

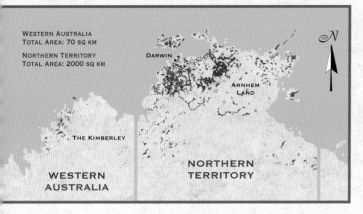

WESTERN AUSTRALIA
TOTAL AREA: 70 SQ KM

NORTHERN TERRITORY
TOTAL AREA: 2000 SQ KM

DARWIN

ARNHEM
LAND

THE KIMBERLEY

NORTHERN
TERRITORY

WESTERN
AUSTRALIA

N

WESTERN AUSTRALIA AND NORTHERN TERRITORY

Western Australia's Kimberley Region contains more than 1500 small, scattered patches of rainforest, ranging from tall stands along rivers to vine thickets in drier areas. Until recent times, the Aboriginal people regularly burned the country around these rainforest fragments, preventing the damage which today results from "hot" fires which combust accumulated ground-litter. Cattle and feral pigs also damage these fragile areas of forest, which are the habitat of more than 200 species of land snails, some found nowhere else but in their one small patch of forest.

In Arnhem Land, in the Northern Territory, streams may be bordered by tall evergreen rainforest, and vine forests occur on sandstone country and lowland river plains. Some rainforest grows around springs which rise on the coastal plain and along the Arnhem Land escarpment. These rainforest patches are generally smaller than five hectares in extent and are easily damaged by feral pigs, cattle and water buffalo. Scenic areas, which contain waterfalls and lush greenery are also the focus of considerable human activity. About seven per cent of Territory rainforest is in Conservation Reserves, and nearly ninety per cent is on Aboriginal land.

Florence Falls, Litchfield National Park, Northern Territory

A walkway such as this one in Victoria's rainforest makes an exciting view of the canopy possible

THE JOURNEY INTO RAINFOREST

Part of our nature as human beings is to dream — to dream of travelling far, exploring strange places, meeting alien life forms, becoming wiser and more content as new experiences become part of us. The rainforest offers all of us the opportunity to realise our dreams. It is unlike any other place on Earth and to many people is a mysterious and exotic world. However, to know the rainforest is to fall under its spell. Enter its galleries with an open mind, eyes and ears tuned to the life of the forest, and soon you will find that its natural rhythms have become part of your being.

There is excitement and adventure in Australia's rainforests but, above all, there is relief from stress, as well as peace and tranquillity.

Previous pages: Rainforest, Otway Ranges National Park, Victoria
Left: Wooroonooran National Park, North Queensland